THIS JOURNAL BELONGS TO:

"If only you knew the magnificence of the numbers 3, 6, and 9, then you would have the key to the universe."

- NIKOLA TESLA

I, ______________________, vow on this day to commit to my personal growth. I promise to fill this journal out until it is completed, and I promise to love and embrace myself fully throughout the process. I understand that life is ever-changing, and I will not forget that I hold the wisdom and strength to overcome all obstacles. I am grateful for this magical moment and look forward to unraveling the days ahead.

____

SIGNATURE

____

START DATE

____

COMPLETION DATE

We're here because of you.

When you're supporting our business, you're supporting a dream.

Share a picture or video of your journal on **TikTok** or **Instagram** for **20% OFF your next purchase!**

DM @zenfulnote or email keila@zenfulnote.com with your video link to receive your special discount.

Let's vibe!

Learn about The 369 Method, Manifesting Techniques, Shadow Work, and more. Follow our @zenfulnote social channels.

@zenfulnote

This is your 369 Journal.

You will use it to keep track of your goals, take inspired action daily, and align with the frequencies of the universe.

Pay attention to your energy and watch it transmit abundance, success, and creativity in your life.

Ready? *Let us begin.*

WHAT IS THE 3 6 9 METHOD?

The 369 method is a practice that aligns with the law of attraction. This method combines significant numerical patterns with the power of your mind to manifest your dreams into reality. There is a numerical reasoning behind this spiritual practice. Frequency is an important factor in manifestation practices, and tapping into these numbers gives you the frequency to turn your dreams into reality.

By using this simple but powerful method, you will plant seeds of thought and belief into your subconscious mind. Once you complete this journal, you will look back with immense gratitude and awe as you realize how capable you are at shifting your reality. This is the most powerful manifestation technique, but it cannot be done without inspired action. With consistent belief and action, you will accomplish what you once thought was impossible. With the 3 6 9 method, you will be able to manifest anything- from wealth, positive relationships, friendships, good health, and so on.

Nikola Tesla was known for his extraordinary inventions that we use every day and he believed that if you integrate **3 6 9** into your daily routine, you can unlock your true potential.

Our nature is based on a hidden mathematical system. Everything in nature is geometrical and follows geometrical patterns. We can see these patterns present in nature in the form of the Golden Ratio and the Fibonacci Sequence. Multi-cellular organisms are formed by the division of a single cell. A single cell divides into 2 cells, and as the division continues a multicellular organism is formed.

In vortex math there is a pattern that repeats itself infinitely: 1, 2, 4, 8, 7, 5, and so on 1, 2, 4, 8, 7, 5, 1, 2, 4, 8, 7, 5, 1, 2, 4. Note that the numbers 3, 6, and 9 are absent in this pattern. Scientist Marko Rodin believes these numbers represent a vector from the third to fourth dimension which he calls a "flux field." This field is a higher dimensional energy that influences the energy circuit of the other six points. It is believed that this is the secret key to free energy, something which Nikola Tesla mastered and dedicated his life to.

Nikola's theory is that 1 2 4 8 7 and 5 represent the physical world in which we live. Conversely, the numbers 3 6, and 9 represent a vector into the 4th dimension. Thus, Nikola Tesla believed that the numbers 3, 6, and 9 unlocked a realm that we cannot see, a higher dimensional world.

WHY 3 6 9?

The significance of the numbers **3 6 9**

The numbers 3, 6, and 9 show up in the universe and in nature consistently in a very significant way.

If we take a perfect circle's digital root we are left with:
3 + 6 + 0 = 9

A circle has 360 degrees.

If you split a circle in half you get 180 degrees.
The digital root of 180 = 9
180/2 = 90
The digital root of 90 is 9.

If you split 90 in half you get 45.
The digital root of 45 is 9
If you split 45 in half you get 22.5
2 + 2 + 5 = 9

If you split 22.5 in half you get 11.25
1 + 1 + 2 + 5 = 9

As you go on the result is **always** 9.

THE SCIENCE OF HABIT

The science of building habits is an important aspect of human behavior. It has been studied by psychologists, sociologists, and economists, as well as other behavioral scientists. Habits are a part of our everyday lives and have a powerful influence on our behavior and our lives. A habit is a behavior that is repeated so often that it becomes automatic and is difficult to change. Habits can be both helpful or harmful and can be either conscious or subconscious. Habits are formed through the process of classical conditioning, which is when a behavior is associated with a particular stimulus. This is usually done through repetition and reinforcement. The prefrontal cortex is responsible for the conscious effort to create new habits or change existing ones. Building new habits require both the repetition of the behavior and the conscious effort to initiate the behavior.

The 369 method is so powerful because it helps you decondition harmful subconscious habits and replace them with new, positive ones through repetitive scripting and consistent goal-oriented actions. By using this journal daily, you will cognitively restructure your beliefs about your own potential. In turn, this will affect how you perform throughout the day positively.

EVERYTHING IN LIFE IS VIBRATION

Vibration is an essential concept in life because it is the fundamental building block of the universe. Every single thing in the universe is made up of energy, which is constantly vibrating. We can see this in the way that we experience the world and all of the things in it. From the air we breathe and the ground we walk on, to the stars in the sky and the food we eat, everything is vibrating at its own unique frequency. The concept of vibration is not new and has been studied for centuries by various cultures and civilizations. Ancient Greeks believed that the world was made up of four elements: earth, air, fire, and water. They believed that these elements were in constant vibration and that they interacted with each other and influenced the world around them.

In the East, Chinese Taoists and Indian Yogis believed that everything in the universe was connected by a subtle energy called Prana or Qi, and that this energy was in constant vibration.

Today, scientists understand that vibration is a fundamental part of the universe. Everything in the universe is made up of atoms, which are constantly vibrating and emitting energy. This energy, called the electromagnetic spectrum, is responsible for the way we experience the world around us.

WRITING HELPS YOUR BRAIN GROW

Writing something down repeatedly is a proven technique to help our brains grow and manifest things in our lives. It is a form of cognitive training that helps us to focus our attention and create better neural pathways in our brains. When we write something down repeatedly, we are engaging in a form of repetitive learning, which helps us to better remember and understand the information.
Writing something down also helps us to create a plan and take steps to achieve the goal. By focusing our attention on the goal, we create a vision for ourselves that can be seen and understood, making it easier to stay motivated and on track.
The power of writing something down repeatedly lies in its ability to form new neural pathways in our brains. When we write something down, it is a way of focusing our attention on the goal, allowing us to better remember and comprehend the information, and to create a plan that will help us to achieve it. By repeatedly writing our goal down, we are reinforcing the neural pathways associated with the goal, making it easier for us to achieve it. This type of cognitive training also helps us to manifest our desired outcome. When we focus our attention on the goal, it becomes easier for us to visualize it and take the necessary steps to make it happen!

HOW TO USE THIS BOOK

How do I do the 369 method?
Write your desire down 3 times in the morning, 6 times in the afternoon, and 9 times in the evening. Then, take inspired action during the day.

Do I start with the affirmations or actions?
Please start by writing down your desire/affirmations first. This will help prime your subconscious and build emotions associated with your desire.

Do I have to write the same thing down for the entire book?
No, you do not. You can change the way you phrase your desire, and you can change your desire entirely. Throughout your journey, new ideas will come and you may gain more clarity about what you want in life. Because of this, your affirmation can change.

Can I manifest multiple things at once?
Yes, you can. If you would like to manifest 2 things, use 2 separate pages so you can keep your intentions organized.

Do I have to write my desire down at certain times of the day?
No, you do not. It is recommended to use the morning, afternoon, and evening. However, it is acceptable and still effective if you choose to write down all of the affirmations at once in the morning.

What can I pair with this book to help me achieve my goals?
Listen to sound healing frequencies, repeat positive affirmations, meditate, exercise, and incorporate movement throughout your day. Listening to an inspiring song while you are writing your desires will amplify your practice even more.

What if I miss a day?
It is OK if you miss a day. You can always pick it back up again.

3 6 9 METHOD

How to manifest with the **3 6 9** method.

Write your desire *3* times in the *morning.*

✦ ✦ ✦

Write your desire *6* times in the *afternoon.*

✦ ✦ ✦ ✦ ✦ ✦

Write your desire *9* times in the *evening.*

✦ ✦ ✦ ✦ ✦ ✦ ✦ ✦ ✦

"I am so happy and my heart is so full now that I have $10,000 extra in my checking account."

@zenfulnote

3 6 9 METHOD EXAMPLE

morning

3

I got a promotion this week
I got a promotion this week
I got a promotion this week

afternoon

6

I got a promotion this week
I got a promotion this week
I got a promotion this week
I got a promotion this week
I got a promotion this week
I got a promotion this week

evening

9

I got a promotion this week
I got a promotion this week
I got a promotion this week
I got a promotion this week
I got a promotion this week
I got a promotion this week
I got a promotion this week
I got a promotion this week
I got a promotion this week

TIPS TO AMPLIFY YOUR PRACTICE

- Pick 1 thing you would like to manifest into your life. Write what you desire 3,6, and 9 times throughout your day. Make sure to be very specific with your affirmation.
- Phrase your affirmations as though you have the thing you're manifesting. For example: "I am so grateful and my heart is so full that I received $1,000 this week."
- Use phrases like "I am", "I have", "I will" in your affirmations. Avoid phrases like "I need" and "I want".
- Use the *Abraham Hicks 17 Second Rule:* After each day of 369 affirmations, take 17 seconds to visualize your desire. How does it feel? Who would you tell the good news to? Surrender to the creative power of your imagination.

30 IDEAS FOR MANIFESTING

ROMANCE

A loving relationship
A new relationship
Marriage
Pregnancy
Good sex life

SPIRITUALITY

Alignment with the universe
High-vibe life
Improving manifestation
A sign from the universe
Complete abundance

HEALTH

Weight loss
Clear skin
Toned body
Better sleep
Good health

FINANCIAL

Passive income streams
Financial independence
New ways of earning
Unexpected money
Being debt-free

CAREER

New job
Promotion
Better working
Relationships
Career change
Your own business

RELATIONSHIPS

New friendships
A best friend
Better family dynamics
Unexpected kindness

Affirmations for *Abundance*

I don't chase, I attract. What belongs to me will simply find me

I can do anything I set my mind to

The world is rigged in my favor

I accept myself for who I am

I have the power to create change

I am grateful for all that I have

I achieve my goals by taking action

There is abundance all around me

Today is the best day of my life

I focus on the positive

I deeply love and accept myself

Affirmations for *Relationships*

I am ready for a loving relationship

I am grateful for my past relationships

I've done the work and now I'm ready for love

I release all separation and judgment

I'm willing to view others with love and kindness

Relationships are mirrors of my ability to love myself

Loving myself allows me to love others

My heart is prepared to receive love

Affirmations for *Career & wealth*

Money comes to me easily and effortlessly

Wealth constantly flows into my life

My life is full of abundance

Money flows freely to me

Money is the easiest thing for me to manifest

I am working hard towards my ambitions

I am capable of reaching the goals I set for my future self

I am a magnet for success

My work makes a difference

I am creating the life I want

Affirmations for H*ealth*

I give myself permission to heal

I appreciate and love my body

I radiate good health

I am calm and at peace

I have all the energy I need to
accomplish my goals

My body is healed, restored, and filled with energy

I am healing

My tension is melting away

I am doing my best and that is enough

SELF-CARE MENU 

Loving actions you can do to take care of yourself.

Take a few deep breaths............. 5 Min

Stretch your body........................ 5 Min

Spend some time in the sun......5 Min

Read a chapter of a book............10 Min

Meditate.. 10 Min

Burn your favorite candle..........10 Min

Take a walk outside.......................30 Min

Put on music and dance away...30 Min

Read or write poetry.................... 30 Min

@zenfulnote

YOUR MIND

IS A POWERFUL

THING. WHEN YOU

FILL IT WITH

POSITIVE THOUGHTS

YOUR LIFE WILL

START TO

CHANGE.

THE 369 JOURNAL

what actions are you taking today?

to-do

- [ ]
- [ ]
- [ ]
- [ ]
- [ ]
- [ ]
- [ ]
- [ ]

How will I show myself love today?

morning

3

afternoon

6

evening

9

what actions are you taking today?

to-do

- [ ]
- [ ]
- [ ]
- [ ]
- [ ]
- [ ]
- [ ]
- [ ]

How will I show myself love today?

morning

3

afternoon

6

evening

9

what actions are you taking today?

to-do

- [ ]
- [ ]
- [ ]
- [ ]
- [ ]
- [ ]
- [ ]
- [ ]

How will I show myself love today?

morning

3

afternoon

6

evening

9

what actions are you taking today?

to-do

- [ ]
- [ ]
- [ ]
- [ ]
- [ ]
- [ ]
- [ ]
- [ ]

How will I show myself love today?

morning

3

afternoon

6

evening

9

what actions are you taking today?

to-do

- [ ]
- [ ]
- [ ]
- [ ]
- [ ]
- [ ]
- [ ]
- [ ]

How will I show myself love today?

morning

3

afternoon

6

evening

9

what actions are you taking today?

to-do

- [ ]
- [ ]
- [ ]
- [ ]
- [ ]
- [ ]
- [ ]
- [ ]

How will I show myself love today?

morning

3

afternoon

6

evening

9

what actions are you taking today?

to-do

- [ ]
- [ ]
- [ ]
- [ ]
- [ ]
- [ ]
- [ ]
- [ ]

How will I show myself love today?

morning

3

afternoon

6

evening

9

what actions are you taking today?

to-do

- [ ]
- [ ]
- [ ]
- [ ]
- [ ]
- [ ]
- [ ]
- [ ]

How will I show myself love today?

morning

3

afternoon

6

evening

9

what actions are you taking today?

to-do

- [ ]
- [ ]
- [ ]
- [ ]
- [ ]
- [ ]
- [ ]
- [ ]

How will I show myself love today?

morning

3

afternoon

6

evening

9

what actions are you taking today?

to-do

- [ ]
- [ ]
- [ ]
- [ ]
- [ ]
- [ ]
- [ ]
- [ ]

How will I show myself love today?

morning

3

afternoon

6

evening

9

what actions are you taking today?

to-do

- [ ]
- [ ]
- [ ]
- [ ]
- [ ]
- [ ]
- [ ]
- [ ]

How will I show myself love today?

morning

3

afternoon

6

evening

9

what actions are you taking today?

to-do

- [ ]
- [ ]
- [ ]
- [ ]
- [ ]
- [ ]
- [ ]
- [ ]

How will I show myself love today?

morning

3

afternoon

6

evening

9

what actions are you taking today?

to-do

- [ ]
- [ ]
- [ ]
- [ ]
- [ ]
- [ ]
- [ ]
- [ ]

How will I show myself love today?

morning

3

afternoon

6

evening

9

what actions are you taking today?

to-do

- [ ]
- [ ]
- [ ]
- [ ]
- [ ]
- [ ]
- [ ]
- [ ]

How will I show myself love today?

morning

3

afternoon

6

evening

9

what actions are you taking today?

to-do

- [ ]
- [ ]
- [ ]
- [ ]
- [ ]
- [ ]
- [ ]
- [ ]

How will I show myself love today?

morning

3

afternoon

6

evening

9

what actions are you taking today?

to-do

- [ ]
- [ ]
- [ ]
- [ ]
- [ ]
- [ ]
- [ ]
- [ ]

How will I show myself love today?

morning

3

afternoon

6

evening

9

what actions are you taking today?

to-do

- [ ]
- [ ]
- [ ]
- [ ]
- [ ]
- [ ]
- [ ]
- [ ]

How will I show myself love today?

morning

3

afternoon

6

evening

9

what actions are you taking today?

to-do

- [ ]
- [ ]
- [ ]
- [ ]
- [ ]
- [ ]
- [ ]
- [ ]

How will I show myself love today?

morning

3

afternoon

6

evening

9

what actions are you taking today?

to-do

- [ ]
- [ ]
- [ ]
- [ ]
- [ ]
- [ ]
- [ ]
- [ ]

How will I show myself love today?

morning

3

afternoon

6

evening

9

what actions are you taking today?

to-do

- [ ]
- [ ]
- [ ]
- [ ]
- [ ]
- [ ]
- [ ]
- [ ]

How will I show myself love today?

morning

3

afternoon

6

evening

9

what actions are you taking today?

to-do

- [ ]
- [ ]
- [ ]
- [ ]
- [ ]
- [ ]
- [ ]
- [ ]

How will I show myself love today?

morning

3

afternoon

6

evening

9

what actions are you taking today?

to-do

- [ ]
- [ ]
- [ ]
- [ ]
- [ ]
- [ ]
- [ ]
- [ ]

How will I show myself love today?

morning

3

afternoon

6

evening

9

what actions are you taking today?

to-do

- [ ]
- [ ]
- [ ]
- [ ]
- [ ]
- [ ]
- [ ]
- [ ]

How will I show myself love today?

morning

3

afternoon

6

evening

9

what actions are you taking today?

to-do

- [ ]
- [ ]
- [ ]
- [ ]
- [ ]
- [ ]
- [ ]
- [ ]

How will I show myself love today?

morning

3

afternoon

6

evening

9

what actions are you taking today?

to-do

- [ ]
- [ ]
- [ ]
- [ ]
- [ ]
- [ ]
- [ ]
- [ ]

How will I show myself love today?

morning

3

afternoon

6

evening

9

what actions are you taking today?

to-do

- [ ]
- [ ]
- [ ]
- [ ]
- [ ]
- [ ]
- [ ]
- [ ]

How will I show myself love today?

morning

3

afternoon

6

evening

9

what actions are you taking today?

to-do

- [ ]
- [ ]
- [ ]
- [ ]
- [ ]
- [ ]
- [ ]
- [ ]

How will I show myself love today?

morning

3

afternoon

6

evening

9

what actions are you taking today?

to-do

- [ ]
- [ ]
- [ ]
- [ ]
- [ ]
- [ ]
- [ ]
- [ ]

How will I show myself love today?

morning

3

afternoon

6

evening

9

what actions are you taking today?

to-do

- [ ]
- [ ]
- [ ]
- [ ]
- [ ]
- [ ]
- [ ]
- [ ]

How will I show myself love today?

morning

3

afternoon

6

evening

9

what actions are you taking today?

to-do

- [ ]
- [ ]
- [ ]
- [ ]
- [ ]
- [ ]
- [ ]
- [ ]

How will I show myself love today?

morning

3

afternoon

6

evening

9

what actions are you taking today?

to-do

- [ ]
- [ ]
- [ ]
- [ ]
- [ ]
- [ ]
- [ ]
- [ ]

How will I show myself love today?

what actions are you taking today?

to-do

- [ ]
- [ ]
- [ ]
- [ ]
- [ ]
- [ ]
- [ ]
- [ ]

How will I show myself love today?

what actions are you taking today?

to-do

- [ ]
- [ ]
- [ ]
- [ ]
- [ ]
- [ ]
- [ ]
- [ ]

How will I show myself love today?

morning

3

afternoon

6

evening

9

what actions are you taking today?

to-do

- [ ]
- [ ]
- [ ]
- [ ]
- [ ]
- [ ]
- [ ]
- [ ]

How will I show myself love today?

morning

3

afternoon

6

evening

9

what actions are you taking today?

to-do

- [ ]
- [ ]
- [ ]
- [ ]
- [ ]
- [ ]
- [ ]
- [ ]

How will I show myself love today?

morning

3

afternoon

6

evening

9

what actions are you taking today?

to-do

- [ ]
- [ ]
- [ ]
- [ ]
- [ ]
- [ ]
- [ ]
- [ ]

How will I show myself love today?

morning

3

afternoon

6

evening

9

what actions are you taking today?

to-do

- [ ]
- [ ]
- [ ]
- [ ]
- [ ]
- [ ]
- [ ]
- [ ]

How will I show myself love today?

morning

3

afternoon

6

evening

9

what actions are you taking today?

to-do

- [ ]
- [ ]
- [ ]
- [ ]
- [ ]
- [ ]
- [ ]
- [ ]

How will I show myself love today?

morning

3

afternoon

6

evening

9

what actions are you taking today?

to-do

- [ ]
- [ ]
- [ ]
- [ ]
- [ ]
- [ ]
- [ ]
- [ ]

How will I show myself love today?

morning

3

afternoon

6

evening

9

what actions are you taking today?

to-do

- [ ]
- [ ]
- [ ]
- [ ]
- [ ]
- [ ]
- [ ]
- [ ]

How will I show myself love today?

morning

3

afternoon

6

evening

9

what actions are you taking today?

to-do

- [ ]
- [ ]
- [ ]
- [ ]
- [ ]
- [ ]
- [ ]
- [ ]

How will I show myself love today?

morning

3

afternoon

6

evening

9

what actions are you taking today?

to-do

- [ ]
- [ ]
- [ ]
- [ ]
- [ ]
- [ ]
- [ ]
- [ ]

How will I show myself love today?

morning

3

afternoon

6

evening

9

what actions are you taking today?

to-do

- [ ]
- [ ]
- [ ]
- [ ]
- [ ]
- [ ]
- [ ]
- [ ]

How will I show myself love today?

morning

3

afternoon

6

evening

9

what actions are you taking today?

to-do

- [ ]
- [ ]
- [ ]
- [ ]
- [ ]
- [ ]
- [ ]
- [ ]

How will I show myself love today?

morning

3

afternoon

6

evening

9

 DATE: / /

what actions are you taking today?

to-do

- [ ]
- [ ]
- [ ]
- [ ]
- [ ]
- [ ]
- [ ]
- [ ]

How will I show myself love today?

morning

3

afternoon

6

evening

9

what actions are you taking today?

to-do

- [ ]
- [ ]
- [ ]
- [ ]
- [ ]
- [ ]
- [ ]
- [ ]

How will I show myself love today?

morning

3

afternoon

6

evening

9

what actions are you taking today?

to-do

- [ ]
- [ ]
- [ ]
- [ ]
- [ ]
- [ ]
- [ ]
- [ ]

How will I show myself love today?

morning

3

afternoon

6

evening

9

what actions are you taking today?

to-do

- [ ]
- [ ]
- [ ]
- [ ]
- [ ]
- [ ]
- [ ]
- [ ]

How will I show myself love today?

morning

3

afternoon

6

evening

9

what actions are you taking today?

to-do

- [ ]
- [ ]
- [ ]
- [ ]
- [ ]
- [ ]
- [ ]
- [ ]

How will I show myself love today?

morning

3

afternoon

6

evening

9

what actions are you taking today?

to-do

- [ ]
- [ ]
- [ ]
- [ ]
- [ ]
- [ ]
- [ ]
- [ ]

How will I show myself love today?

morning

3

afternoon

6

evening

9

what actions are you taking today?

to-do

- [ ]
- [ ]
- [ ]
- [ ]
- [ ]
- [ ]
- [ ]
- [ ]

How will I show myself love today?

morning

3

afternoon

6

evening

9

what actions are you taking today?

to-do

- [ ]
- [ ]
- [ ]
- [ ]
- [ ]
- [ ]
- [ ]
- [ ]

How will I show myself love today?

morning

3

afternoon

6

evening

9

what actions are you taking today?

to-do

- [ ]
- [ ]
- [ ]
- [ ]
- [ ]
- [ ]
- [ ]
- [ ]

How will I show myself love today?

morning

3

afternoon

6

evening

9

what actions are you taking today?

to-do

- [ ]
- [ ]
- [ ]
- [ ]
- [ ]
- [ ]
- [ ]
- [ ]

How will I show myself love today?

morning

3

afternoon

6

evening

9

what actions are you taking today?

to-do

- [ ]
- [ ]
- [ ]
- [ ]
- [ ]
- [ ]
- [ ]
- [ ]

How will I show myself love today?

morning

3

afternoon

6

evening

9

what actions are you taking today?

to-do

- [ ]
- [ ]
- [ ]
- [ ]
- [ ]
- [ ]
- [ ]
- [ ]

How will I show myself love today?

morning

3

afternoon

6

evening

9

what actions are you taking today?

to-do

- [ ]
- [ ]
- [ ]
- [ ]
- [ ]
- [ ]
- [ ]
- [ ]

How will I show myself love today?

morning

3

afternoon

6

evening

9

what actions are you taking today?

to-do

- [ ]
- [ ]
- [ ]
- [ ]
- [ ]
- [ ]
- [ ]
- [ ]

How will I show myself love today?

morning

3

afternoon

6

evening

9

what actions are you taking today?

to-do

- [ ]
- [ ]
- [ ]
- [ ]
- [ ]
- [ ]
- [ ]
- [ ]

How will I show myself love today?

morning

3

afternoon

6

evening

9

what actions are you taking today?

to-do

- [ ]
- [ ]
- [ ]
- [ ]
- [ ]
- [ ]
- [ ]
- [ ]

How will I show myself love today?

morning

3

afternoon

6

evening

9

what actions are you taking today?

to-do

- [ ]
- [ ]
- [ ]
- [ ]
- [ ]
- [ ]
- [ ]
- [ ]

How will I show myself love today?

morning

3

afternoon

6

evening

9

what actions are you taking today?

to-do

- [ ]
- [ ]
- [ ]
- [ ]
- [ ]
- [ ]
- [ ]
- [ ]

How will I show myself love today?

morning

3

afternoon

6

evening

9

what actions are you taking today?

to-do

- [ ]
- [ ]
- [ ]
- [ ]
- [ ]
- [ ]
- [ ]
- [ ]

How will I show myself love today?

morning

3

afternoon

6

evening

9

what actions are you taking today?

to-do

- [ ]
- [ ]
- [ ]
- [ ]
- [ ]
- [ ]
- [ ]
- [ ]

How will I show myself love today?

morning

3

afternoon

6

evening

9

what actions are you taking today?

to-do

- [ ]
- [ ]
- [ ]
- [ ]
- [ ]
- [ ]
- [ ]
- [ ]

How will I show myself love today?

morning

3

afternoon

6

evening

9

what actions are you taking today?

to-do

- [ ]
- [ ]
- [ ]
- [ ]
- [ ]
- [ ]
- [ ]
- [ ]

How will I show myself love today?

morning

3

afternoon

6

evening

9

what actions are you taking today?

to-do

- [ ]
- [ ]
- [ ]
- [ ]
- [ ]
- [ ]
- [ ]
- [ ]

How will I show myself love today?

morning

3

afternoon

6

evening

9

what actions are you taking today?

to-do

- [ ]
- [ ]
- [ ]
- [ ]
- [ ]
- [ ]
- [ ]
- [ ]

How will I show myself love today?

morning

3

afternoon

6

evening

9

what actions are you taking today?

to-do

- [ ]
- [ ]
- [ ]
- [ ]
- [ ]
- [ ]
- [ ]
- [ ]

How will I show myself love today?

morning

3

afternoon

6

evening

9

what actions are you taking today?

to-do

- [ ]
- [ ]
- [ ]
- [ ]
- [ ]
- [ ]
- [ ]
- [ ]

How will I show myself love today?

morning

3

afternoon

6

evening

9

what actions are you taking today?

to-do

- [ ]
- [ ]
- [ ]
- [ ]
- [ ]
- [ ]
- [ ]
- [ ]

How will I show myself love today?

morning

3

afternoon

6

evening

9

what actions are you taking today?

to-do

- [ ]
- [ ]
- [ ]
- [ ]
- [ ]
- [ ]
- [ ]
- [ ]

How will I show myself love today?

morning

3

afternoon

6

evening

9

what actions are you taking today?

to-do

- [ ]
- [ ]
- [ ]
- [ ]
- [ ]
- [ ]
- [ ]
- [ ]

How will I show myself love today?

morning

3

afternoon

6

evening

9

what actions are you taking today?

to-do

- [ ]
- [ ]
- [ ]
- [ ]
- [ ]
- [ ]
- [ ]
- [ ]

How will I show myself love today?

morning

3

afternoon

6

evening

9

DATE: / /

what actions are you taking today?

to-do

- [ ]
- [ ]
- [ ]
- [ ]
- [ ]
- [ ]
- [ ]
- [ ]

How will I show myself love today?

morning

3

afternoon

6

evening

9

what actions are you taking today?

to-do

- [ ]
- [ ]
- [ ]
- [ ]
- [ ]
- [ ]
- [ ]
- [ ]

How will I show myself love today?

morning

3

afternoon

6

evening

9

what actions are you taking today?

to-do

- [ ]
- [ ]
- [ ]
- [ ]
- [ ]
- [ ]
- [ ]
- [ ]

How will I show myself love today?

morning

3

afternoon

6

evening

9

what actions are you taking today?

to-do

- [ ]
- [ ]
- [ ]
- [ ]
- [ ]
- [ ]
- [ ]
- [ ]

How will I show myself love today?

morning

3

afternoon

6

evening

9

what actions are you taking today?

to-do

- [ ]
- [ ]
- [ ]
- [ ]
- [ ]
- [ ]
- [ ]
- [ ]

How will I show myself love today?

morning

3

afternoon

6

evening

9

what actions are you taking today?

to-do

- [ ]
- [ ]
- [ ]
- [ ]
- [ ]
- [ ]
- [ ]
- [ ]

How will I show myself love today?

morning

3

afternoon

6

evening

9

what actions are you taking today?

to-do

- [ ]
- [ ]
- [ ]
- [ ]
- [ ]
- [ ]
- [ ]
- [ ]

How will I show myself love today?

morning

3

afternoon

6

evening

9

what actions are you taking today?

to-do

- [ ]
- [ ]
- [ ]
- [ ]
- [ ]
- [ ]
- [ ]
- [ ]

How will I show myself love today?

morning

3

afternoon

6

evening

9

what actions are you taking today?

to-do

- [ ]
- [ ]
- [ ]
- [ ]
- [ ]
- [ ]
- [ ]
- [ ]

How will I show myself love today?

morning

3

afternoon

6

evening

9

what actions are you taking today?

to-do

- [ ]
- [ ]
- [ ]
- [ ]
- [ ]
- [ ]
- [ ]
- [ ]

How will I show myself love today?

morning

3

afternoon

6

evening

9

what actions are you taking today?

to-do

- [ ]
- [ ]
- [ ]
- [ ]
- [ ]
- [ ]
- [ ]
- [ ]

How will I show myself love today?

morning

3

afternoon

6

evening

9

what actions are you taking today?

to-do

- [ ]
- [ ]
- [ ]
- [ ]
- [ ]
- [ ]
- [ ]
- [ ]

How will I show myself love today?

morning

3

afternoon

6

evening

9

what actions are you taking today?

to-do

- [ ]
- [ ]
- [ ]
- [ ]
- [ ]
- [ ]
- [ ]
- [ ]

How will I show myself love today?

morning

3

afternoon

6

evening

9

what actions are you taking today?

to-do

- [ ]
- [ ]
- [ ]
- [ ]
- [ ]
- [ ]
- [ ]
- [ ]

How will I show myself love today?

morning

3

afternoon

6

evening

9

what actions are you taking today?

to-do

- [ ]
- [ ]
- [ ]
- [ ]
- [ ]
- [ ]
- [ ]
- [ ]

How will I show myself love today?

morning

3

afternoon

6

evening

9

what actions are you taking today?

to-do

- [ ]
- [ ]
- [ ]
- [ ]
- [ ]
- [ ]
- [ ]
- [ ]

How will I show myself love today?

morning

3

afternoon

6

evening

9

what actions are you taking today?

to-do

- [ ]
- [ ]
- [ ]
- [ ]
- [ ]
- [ ]
- [ ]
- [ ]

How will I show myself love today?

morning

3

afternoon

6

evening

9

what actions are you taking today?

to-do

- [ ]
- [ ]
- [ ]
- [ ]
- [ ]
- [ ]
- [ ]
- [ ]

How will I show myself love today?

morning

3

afternoon

6

evening

9

what actions are you taking today?

to-do

- [ ]
- [ ]
- [ ]
- [ ]
- [ ]
- [ ]
- [ ]
- [ ]

How will I show myself love today?

morning

3

afternoon

6

evening

9

what actions are you taking today?

to-do

- [ ]
- [ ]
- [ ]
- [ ]
- [ ]
- [ ]
- [ ]
- [ ]

How will I show myself love today?